Enhancing Army S&T
Vol. II: The Future

John W. Lyons and Richard Chait

Center for Technology and National Security Policy
National Defense University

March 2012

The views expressed in this article are those of the authors and do not reflect the official policy or position of the National Defense University, Department of Defense, or U.S. Government. All information and sources for this paper were drawn from unclassified materials.

John W. Lyons is a Distinguished Research Fellow at the Center for Technology and National Security Policy (CTNSP), National Defense University. He was previously director of the Army Research Laboratory and director of the National Institute of Standards and Technology. Dr. Lyons received his PhD in physical chemistry from Washington University and a BA in chemistry from Harvard.

Richard Chait is a Distinguished Research Fellow at CTNSP. He was previously Chief Scientist, Army Material Command; and Director, Army Research and Laboratory Management. Dr. Chait received his PhD in solid state science from Syracuse University and a BS from Rensselaer Polytechnic Institute.

Acknowledgements: The authors wish to acknowledge the support of Dr. Marilyn Freeman, Deputy Assistant Secretary of the Army for Research and Technology; and Dr. Linton Wells, Director of CTNSP. The authors are grateful for helpful comments from Dr. Steve Ramberg and Dr. James J. Valdes.

CONTENTS

LIST OF FIGURES

Executive Summary

The Army team at the Center for Technology and National Security Policy has been doing technology studies for the Deputy Assistant Secretary of the Army for Research and Technology since 2003. In 2007 we published *Enhancing Army S&T: Lessons Learned From Project Hindsight Revisited,* which we refer to here as Vol. I. That publication was a summary of critical technology contributions to the development of four successful Army warfighting systems. Since then, we have completed a number of studies of important aspects of the Army science and technology (S&T) program with an emphasis on the Army laboratories. In the present paper, Vol. II, we integrate the findings of these studies and make recommendations after each chapter, as well as in a separate final chapter.

Chapter I of this volume is an introduction, and Chapter II offers an updated view of the work discussed in Vol. I with an emphasis on the relative roles played by the Army laboratories and the contractors that manufactured the systems. The close collaboration between the two groups was judged by us to be the key to the successful outcomes. Both the Army laboratories and the technical personnel at the contractors were essential—without either group the work would have cost more, taken more time, and might well have failed. We believe the collaboration was the result of the efforts of the mid-level managers who pressed technologists to work together.

We recommend the Army continue to emphasize the importance of maintaining close working relationships between the laboratories and the technical staffs of Army contractors.

In Chapter III, we discuss the impact of the lack of publicity given to the Army laboratories' work. This lack of publicity has caused some observers to conclude that the laboratories are not significant contributors to the warfighters. This belief in turn has produced recommendations from outside the military to close the laboratories and assign the research to the private sector. We do not agree with the criticism or the recommendation. We discuss two aspects of addressing this problem: the need to maintain high-quality work and the need to provide detailed information about the contributions of the laboratories to all parties concerned—namely, Army senior leadership, officials in the Department of Defense (DOD), the Administration, the Congress, and the general public.

We recommend an aggressive campaign to publicize the technical contributions of the laboratories to the senior leadership of the Army, to other stakeholders, and to the general public.

Chapter IV explores the laboratory quality question. We begin by asserting that the most important asset of a laboratory is its technical staff members and that, therefore, ensuring staff quality should be a top priority of management. We discuss a number of methods for locating and bringing new employees onboard, including use of the Intergovernmental Personnel Act (IPA), post-doctoral appointments, and visiting scientists and engineers.

We recommend extensive use of the IPA, increased use of post-doctoral appointments, and more extended visits by senior scientists and engineers to Army laboratories.

The more than 40 Senior Technical Professional s (ST) in the Arm y represent the best S&T performers the Army has; their presence and in fluence should be m aximized. Our survey of the STs revealed a lack of uniform ity in how they are funded, m anaged, and utilized in providing advice and counsel in program planning.

We recommend the Army support the STs and make better use of their many talents.

A strong program of basic, fundamental research provides the foundation for applied and perhaps shorter term research. It also attracts well-qualified new additions to the staff. Some of the Arm y laboratories do not have enough funding for basic research.

We recommend that each laboratory should h ave a significant level of funding for basic research.

To ensure the laboratory is satisfying the requi rements for quality, relevance, and tim eliness, assessments should be m ade on a regular basis. Discussions should be held with custom ers and stakeholders, and independent external expe rts should assess quality. Com prehensive peer review strengthens the laboratory programs, and the reports help improve the laboratory's image.

We recommend that the Assis tant Secretary of the Ar my for Acquisition, L ogistics, and Technology (ASAALT) require, and each Army laboratory should arrange for, regular quality assessments by independent, external subject matter experts.

Chapter V discusses two reports we issued on the role of technol ogy in stabilization and reconstruction. We surveyed the experien ces of recently return ed soldiers from Iraq. More recently we have con ducted Gedanken Experiments at Fort Bennning to exp lore, with experienced soldiers, various challenges facing the laboratory program s. These experim ents brought together a number of officers and senior non-commissioned officers in combination with Army scientists and engineers and observers from ASAALT and other A rmy organizations. The participants have been enthusiastic about th e experience and are urging that m ore such experiments be carried out.

We recommend that, in defining and shaping research programs, interactions w ith experienced warfighters should increase further.

DOD's Independent Research and Developm ent (IR&D) program supports research by defense contractors in preparation for fu ture contracted work. The Ar my laboratories, for the m ost part, are not tracking this work. Both the Navy and the Air Force have formal programs to learn about the research programs of their contractors. We believe the Army is missing an opportunity here.

We recommend that all of the Army laboratori es undertake more form al interaction with industry's IR&D programs.

The Army has not conducted a long-range technol ogy forecast since 1992. It is tim e to begin a new forecasting effort. A review of current practic es shows that structuring forecasts to look for convergences among a set of sciences or technologi es could lead to breakthroughs in addressing capabilities not currently available to the warfighter.

We recommend ASAALT conduct periodic technology forecasting based on th e concept of convergences.

A trend in conducting research   and developm ent is close,   long-term collaboration am ong laboratories. This collaboration could be betw   een different laboratories or individuals. Institutional collaboration as represented by the Army Research Laboratory (ARL) Collaborative Technology Alliances has been successful in terms of gathering new information and broadening the knowledge of ARL participan ts and consortia m embers. Collaboration is a worldwide phenomenon facilitated by the Internet and fibe r optic links across count ries and across the oceans.

We recommend that the Army continue to use the mechanism of formal collaboration with other companies and universities.

The evolution of com puting has brought us to the point where the high-performance computer (HPC) capability of the past is now found in desktop and laptop devices. Modern HPC machines can conduct the development and testing of desi gn options for products and system s in minutes as opposed to weeks or m onths. In so doing, HPC can greatly shorten the entire innovation process. Industry has dem onstrated this ca pability. The United States—both industry and government—is putting a new emphasis on manufacturing technology, and HPC is a part of it.

We recommend that the Army emphasize the use of HPC in the design and manufacture of its systems.

Numerous studies in the past have demonstrated the role of military research laboratories. In the Army, such studies were done during Base R ealignment and Closure (BRAC) 91 and, m ore recently, in the Arm y Materiel Comm and's creation of the Research, Developm ent, and Engineering Command. New st udies are now underway based on concerns ab out creativity, visibility, and costs. Reporting relationships and access to policymakers are part of these studies.

In Chapter VI w e recommend th at the Army laboratories be man aged as the important component of developing new capabilities for warfighters that they are. The Army should emphasize reporting relationships and the ro le of ASAALT in developing policy affecting the laboratories.

CHAPTER I. INTRODUCTION

In 2007 the Ar my team at the Center for T echnology and National Security Policy (CTNSP) published a summary of a group of studies on the role Arm y technology plays in the development of four successful Army platforms. [1] We refer to the 2007 paper as Vol. I. That publication includes a section on ex cellence at the Army laboratories. We based this section on our extensive interactions with the Army laboratories and Army contractors for the four systems, as well as on our extensive experience with Army technology. Since then, we have continued to study several aspects of the exce llence issue for the Arm y laboratories. In the present volum e, which we have titled V ol. II, we in tegrate these studies into a set of recomm endations for the Assistant Secretary of the Ar my for Acquisition, Logistics, a nd Technology (ASAALT) and for the Deputy Assistant Secretary of the Army for Research an d Technology (DAS(RT). (Herein these two offices will be referred to together as ASAALT).

We note that historically the Army operated arsenals whose purpose was to produce the arms and armaments needed. The arsenals developed a nd then m anufactured the products. Exa mples include the Watertown Arsenal, which supplied the Rodm an guns researched and designed by the commander of the arsenal; the Picatinny Ar senal, which produced both sm all and medium arms and ammunition; and the W atervliet Arsenal, which manufactured large gun barrels. Most of the arsenals no longer focus on production but ra ther are responsible for sustainm ent. Today, two remaining arsenals m anufacture major Army components: Watervliet for gun tubes and Rock Island for gun mounts and a variety of we apons components and systems. Private sector contractors manufacture weapons platform s, including tanks, arm ored personnel carriers, artillery pieces, and wheeled veh icles, such as trucks and High Mobility Multipurpose Wheeled Vehicles (HMMWV). Army laboratories perform much of the up-front research and development (R&D) (basic research through advanced development, categories 6.1, 6.2, and 6.3) in close collaboration with th e contractors selected to m anufacture the products. The Ar my primarily performs the research; the research s taff have clo se ties to the large contractors that produce the products. The contr actors do perfor m some research connected w ith current contracts, and, under the Indepe ndent Research and Development (IR&D) program (see Chapter V), they perform research in preparation for likely future projects.

External critics have proposed privatizing al l of the R&D w ork now done by the A rmy. These proposals have attem pted to sh ift responsibility to private industry. The m otivation for the criticisms appears to in clude three ideas: that the private sector can do the work at lower cost; that private R&D is inherently of higher quality than military R&D; and that, philosophically, the least government, the better. So far, proposals to privatize the laborator ies have not gained traction largely because they are an im portant cog in th e weapons developm ent process. In addition, as noted by Coffey, [2] it is important to m aintain a corps of technical experts in house. In-house expertise protects the military from unsubstantiated claims by contractors and others and thus maintains a "smart buyer" perspective. One of the objectives of this paper is to continue

---

[1] Richard Chait, John Lyons, Duncan Long, and Albert Sciarretta, *Enhancing Army S&T: Lessons from Project Hindsight Revisited* (Washington, DC: Center for Technology and National Security Policy [CTNSP], National Defense University).

[2] Timothy Coffey, *Building the S&E Workforce for 2040, Challenges Facing the Department of Defense,* Defense & Technology Paper 49 (Washington, DC: CTNSP, National Defense University, July 2008).

to emphasize the importance of the Army laboratories and to recomm end improvements in the Army's management of them.

We begin with our initial studies of the role of Army laboratories in the successful development and fielding of four Army warfighting platforms. These studies provide a basis to explore several areas that need more emphasis: reputation (Chapter III), quality (Chapter IV), portfolio planning (Chapter V), and laboratory management (Chapter VI). At the end of each chapter in this paper, we review the conclusions from these studies. The final chapter reviews our conclusions and outlines a set of recommendations for Army senior management to evaluate and implement them.

CHAPTER II. IMPACT OF PREVIOUS SCIENCE AND TECHNOLOGY EFFORTS

One of our first challenges was to exam ine an old (1969) report on evaluating the results of investing R&D dollar s in m ilitary research. That repo rt[3] addresses two m ain issues raised, in part, by the Congress: (1) identifying those management factors that are im portant in ensuring R&D programs are productive and (2) identifying the increase in cost-effectiveness assignable to the Department of Defense's (DOD) investment in R&D. A la rge team drawn from across DOD devoted 4 years to studying 20 weapons system s. The team docum ented the im pact of R&D performed by the military and industry and s howed that contr ibutions by both sectors were essential.

In 2004 ASAALT asked us to reprise the 1969 study for Army weapons system s that had been developed since Project Hindsight. We studied four weapons syst ems: the Abram s main battle tank,[4] Apache attack helicopter, [5] Stinger m issile system, and Javelin missile system. [6] These findings are summarized in Vol. I. Note that th e objective of these studies was to seek out the critical technology events (CTE) that contributed to successes during th e acquisition cycle. We looked for new and innovative concepts, devices , and com ponents that contributed in a significant way to the end system s. We talked—in detail—in person or by telephone to 158 people and discovered 134 CTEs for the four systems. The Abrams report is based on interviews with 60 people who were active in the period of development (roughly from 1971 to 1999). We obtained first-hand information on 55 CTEs that were d eemed critical to the success of the Abrams program. We found that the funding for th ese technical programs came almost entirely from the U.S. Government. Som e CTEs cam e from work sponsored by DOD but executed by industry or academia; a few ca me from laboratorie s of other countries. Many CTEs arose from close collaboration among various contributors from the Government and the private sector. The Army Program Manager Office oversaw the inte gration of the new technologies, and the manufacturer of the tank—Chrys ler/General Dynamics—largely carried out the integration. We found that the presence of highly skilled subjec t matter experts (SME) in the Arm y laboratories over long periods of tim e, as well as necessary speciali zed laboratory equipment and facilities, was critical. The Army laboratories were most deeply involved in those aspects of the Abram s that were unique to ground com bat systems. Finally, all interviewees attributed the successful Army contributions to the supportive and patie nt environment provided by the Army laboratory director and staff.

---

[3] Office of the Director of Defense Research and Engineering, *Project Hindsight, Final Report* (Washington, DC: Office of the DDRE, U.S. Department of Defense, 1969)

[4] Richard Chait, John Lyons, and Duncan Long, *Critical Technology Events in the Development of the Abrams Tank, Project Hindsight Revisited,* Defense & Technology Paper 22 (Washington, DC: CTNSP, National Defense University, December 2005).

[5] Richard Chait, John Lyons, and Duncan Long, *Critical Technology Events in the Development of the Apache Helicopter, Project Hindsight Revisited,* Defense & Technology Paper 26 (Washington, DC: CTNSP, National Defense University, February 2006).

[6] John Lyons, Duncan Long, and Richard Chait, *Critical Technology Events in the Development of the Stinger and Javelin Missile Systems, Project Hindsight Revisited,* Defense & Technology Paper 33 (Washington, DC: CTNSP, National Defense University, July 2006).

Figure 1. The Abrams Main Battle Tank

Figure 2. The Apache AH-64 Attack Helicopter

These conclusions applied to all four weapons system s. For the Abram s and the Apache, the government laboratories contributed the m ost CTEs; for the two m issile systems, industry was dominant. The Army and industry technical staffs provided the vast majority of the CTEs. Most striking to us was the closeness of the coll aboration between government and i ndustry. We attribute these collaborative relationships to the efforts of mid-level managers in the technical departments. Our study revealed the important, indeed essential, nature of the role of the Army laboratories. This conclusion caused us to include in Vol. I a chapter on ensuring excellence at the laboratories. This discussion includes much of what has been discussed in many of our reports and is presented in updated form in the earlier chapters of this paper. Vol. I (p. 97) closes as follows: "…for future success the Army in-house laboratories must be supported by sufficient funding, strong leadership, and top flight technical staff … this will require champions at all levels in the Army and DOD…"

Conclusion:
In Project Hindsight Revisited, we learned the importance of close working relationships between industry scientists and engineers and the Army laboratories and acquisition specialists. Success in fielding major Army platforms is attributable to this teamwork; lack of teamwork would have made successful fielding more costly and less timely.

CHAPTER III. ENHANCING THE REPUTATION OF THE LABORATORIES

One motivation for Vol. II is the impact of the lack of external recognition for Army laboratories on the support for Army laboratories. This lack has led to a perception by m any people that the Army laboratories are not of the sam e quality as the best governm ent or private sector laboratories. We devote space here to explore th e reputation of the Ar my laboratories and what should be done to enhance it.

Many have made comparisons between the Army laboratories and government laboratories, such as the Naval Research Laboratory (NRL), Na tional Institute of Standards and Technology (NIST), Department of Energy (DOE) National L aboratories, and the internal laboratories of the National Institutes of Health (NI H). The comparisons a re usually m ade based on externa l recognition, such as awards, m embership in honor societies such as th e National Academ ies, numbers of refereed publications in top-quality journals, and rare honors such as the Nobel prize. The Army laboratories have not fared well in these comparisons

Most criticisms are *not* based on detailed knowledge of the ac tual accomplishments of the Army laboratories. For exam ple, how m any people know that the first all-purpose digital electronic computer, the Electron ic Numerical Integrator and Com puter (ENIAC), was commissioned by the Army's Ballistic Research Laboratory in 1943 and operated for several years at the Arm y's Aberdeen Proving Ground? The E NIAC is consider ed to be the genesis of m odern digital computing.

It is often the case the Ar my's laboratories have contributed key technical i nnovations in the Army's weapons system s, but these contributi ons disappear from vi ew when contractors incorporate them into finished systems. Most observers have no idea what the Army laboratories' contributions have been. (This is a problem at many central corporate laboratories—their work is subsumed in systems fielded by other parts of the company.) It is only through vigorous public relations work that central laboratories have been recognized, such as the Bell L aboratories, General Electric's laboratories, or IBM's laboratories.[7]

Two things are needed to establish the Arm y laboratories' reputation: outstanding quality and more attention to ex ternal relations. Quality is often indirectly m easured by benchm arking against laboratories generally accepted as leaders in the field. Examples[8] of metrics used are—

- Percentage of technical staff with doctorates
- Number of patents
- Number of refereed papers per technical staff member
- Number of staff elected to membership in the National Academies

---

[7] It should be noted that many companies have shifted the focus of their in-house laboratories from a balanced portfolio of short term and longer term work to focus on current challenges. Some have essentially turned to academia for the longer term fundamental needed for creating breakthroughs.

[8] Edward A. Brown, *Reinventing Government Research and Development: A Status Report on Management Initiatives and Reinvention Efforts at the Army Research Laboratory,* ARL-SR-57, Army Research Laboratory (ARL), Adelphi, MD, August 1998; also see ref. 2.

- Number of staff as senior members of professional societies
- Number of prestigious National and international awards.

These metrics are numerical values and do n ot fully capture the true worth of a laboratory. External peer assessment of the work is a good indicator of quality. Th ese assessments can be performed through formal reviews or sim ply by developing a measure of the general regard in which the technical community holds the laboratories.

It is im portant to provide detailed inform ation about the laboratories to the Army's senior leadership and officials in DOD, the Administration, and Congress, as well as the general public. Attention should also be paid to scientific and professional societies. Everyone should know that the Army's technical community has been responsible for m any of the key technologies in the Army's platforms, such as the arm or on vehicles and individual soldiers, arm aments from very large guns down to pistols and rifles, the long r od penetrators known as the Silver Bullets that played a key role in Desert Storm, and so on.

Leading laboratories, such as the Bell Laboratories, did not become recognized for excellence by chance. They recruited the very best scientists and engineers and gave them considerable leeway in the projects they undertook. They sim ply told them about the m ission of the parent organization and how the laboratories' work s hould support that mission. Strong encouragement of creative exploration into ne w areas of science and engineering produced inventions and innovations that were very useful to the com pany but also had m ajor impacts on society. The business area of the Bell Laboratories, for example, is intimately connected to the population as a whole; every home with a telephone quickly recognized the lab's innovati ons. The Army does not have this intim ate relationship with the pub lic, so calling atten tion to new dev elopments in the Army is more difficult. The public only sees the performance of the military in conflicts but not in the detail necessary to appreciate the underlying technical work.

So, the Bell Laboratories start with an advantage over the Army laboratories. But the former also did a m uch better job of pub licizing their work. Everyone know s that the Bell Laboratories invented the transistor and that the transistor is central to the digital revolution, including communications, computers, and consum er products of all kinds. As noted above, very few people are aware of the role of the Ballis tics Research Laboratory in launching th e age of the modern digital computer. The Army S&T program does not do a very good job of publicizing its contributions. The Army holds the Ar my Science Conference, which is a vehicle for the Army technical community prim arily to talk am ong its m embers and to its contractors. The Army should emphasize the value of talking about its S&T work at external scientific meetings.

External recognition also comes to staff members and the laboratories as a whole through awards and election to distinguished rank in m ajor technical societies and to the National Academ ies. The Army does giv e awards, and these are v ery much appreciated. However, awards from outside organizations have a greater effect and generate publicity outside the Arm y. For example, an electrical engineer is highly honored by his peers if elected to the rank of Fellow of the Institute of Electrical and Elec tronics Engineers. For most scientists and engineers, the m ost prized recognition is election to one of the National Academies. Nomination and election are part of a challenging process that takes a lot of work. It involves s oliciting nominators and references from the existing m embers and en suring they and the academ y membership receive all th e necessary background on the person's accomplishments. The Army laboratories have only rarely attempted election to one of the National Academies; the laboratories currently have two active and one retired members in the National Academy of Engineering. Compare this number to the Bell Laboratories and its successors, which currently have 15 active and 20 retired members. Bell and its successors also have members of the National Academy of Science; we know of no members from the Army.

The Nobel Prize is, for most scientists, the most prestigious of all. The prize receives substantial media attention and instantly confers an aura of excellence on the parent laboratory. We are unaware of any Nobel prizes to the Army laboratories and rarely to any military laboratory— NRL has received one, shared by two of its scientists in 1985; the Air Force has received none; NIH has received five; and NIST has received three. We note that the Army Research Office (ARO) has sponsored research with several Nobelists, but none were employed by the Army. Investment in their work is a measure of the wisdom of the ARO program.

It seems Army public relations programs have not emphasized the R&D work done within the Army. The Army laboratories and the media at higher echelons of the Army must make greater attempts to highlight the Army laboratories in publications. Consider the attention the Defense Advanced Research Projects Agency (DARPA) receives for its work. Granted, DARPA focuses on areas where breakthroughs are likely and does not have to worry about supporting the day-to-day needs of the acquisition community and warfighters. When the Army is at war—as it has been for a decade—the immediate needs from the war zones take the highest priority. One does not hear much about that work, even within the Army. One Army publication disseminated by the Office of the ASAALT, *Army AL&T*, occasionally has contributions from Army technologists, but this journal is not widely circulated. An assessment of the information flow about Army technical work would be useful in preparation for a more ambitious effort in information dissemination.

Strengthening the areas discussed in this report would also enhance the Army laboratories' reputation. The following chapters analyze the quality of Army S&T work, the planning function for the S&T program, the S&T program's impact on warfighting capabilities (see, for example, the discussion in Chapter II), and opportunities to strengthen Army laboratory management. We believe implementing the suggestions in this paper would result in appropriate recognition by prestigious organizations, such as the National Academies.

Conclusion:
The reputation of the Army laboratories can improve by achieving outstanding work quality and communicating the impacts of laboratory contributions on the warfighter to senior Army leadership and the general public.

CHAPTER IV. ENHANCING THE QUALITY OF ARMY LABORATORIES

The most important asset of a la boratory is its pers onnel. An outstanding st aff with appropriate management will produce exceptiona l results. Of course, several other factors en able staff to perform up to their potential, such as resour ces (money, equipment, and facilities), supportive management, champions at high levels, and a challenging mix of research programs.

Supporting the Technical Staff. In Vol. I (see Chapter 4), we discussed issues surrounding hiring, retaining, and prom oting technical staff. Care must be taken to obtain the very best new hires and then to m entor them and see to it that they are sup ported with the necessary resources to do their jobs. The form al hiring process is ponde rous compared to that of the private sector entities competing with the Arm y. Some recruiting progress has been m ade, but it is hard to compete with recruiters who have authority to make a binding offer on the spot. The Government does not work that way. Nonetheless, the m ission of the Ar my laboratories to develop a m eans for soldiers to win has strong ap peal to young people, so some of the best and brightest will opt for the m ilitary laboratories. The strongest m otivators for researchers are challenging assignments, good equipm ent and facilities, agreeable cowo rkers, and o pportunities to publish and attend society meetings. Salary is not the strongest m otivator as long as salary decisions are perceived as fair and com pensation, though less than the private sector in m any instances, is reasonable.

Outstanding performers must be treated better than mediocre or poor perfor mers. The top staff should be rewarded with bonuses and m ore opportunities to travel and to publish. Poor performers should be put on probation and given a detailed work plan that, if fulfilled, could raise their performance rating. Failure to meet the plan should lead to termination. History shows that even when given a personnel policy that requi res such performance distinctions to be made, many mid-level managers tend to give m ostly average or som ewhat above-average ratings because differentiating more strongly could lead to m essy appeals and other adverse personnel actions. Many managers are not willing to put up with these human resources issues.

Managers must show appreciation for the work of their most effective senior technical staff, but they do not always do so. W e recently ex amined the status of the Ar my's corps of Senior Technical Professionals (ST). The STs represen t non-managerial personnel who are perform ing at the highest levels in their fields of scien ce or engineering. The Ar my currently has about 40 such individuals; they have deep knowledge of t heir fields and are a source of technical counsel to managers and younger staff. Yet we found that not all Army STs are utilized as key m embers of their organizations. Som e do not have any base funds and m ust sell their talents to external sponsors every year. T hey are not often involved in planning the technical work of their organizations. Their protocol level in the Army is equivalent to th at of a member of the Senior Executive Service (SES) or of a general officer, but they are not treated as such. The Arm y is wasting their talents and potential. We have offered recommendations to ASAALT to correct this situation.[9]

---

[9] Private communication to the Deputy Assistant Secretary of the Army for Research and Technology, February 2010.

A recent ex ample from the Nation al Security Personnel System (NSPS) is instru ctive. The Congress gave DOD authority to transfer em ployees into a new personnel system (NSPS) that called for tying salary to performance ratings. Automatic step increases that occurred in the old system were dropped. In the new system , average performers would only receive cost of living adjustments; poor perform ers would get no incr eases and would eventually fall behind their contemporaries. This sy stem is sim ilar to that used in many companies in the p rivate sector. However, employee unions com plained that the new system placed employees at the m ercy of arbitrary decisions by supervisors. They made such strong protests that eventually the experiment at DOD was cancelled, and NSPS no longer exists. However, the Congress directed[10] DOD to create a new perform ance management system designed to reward high perfor mers yet avoid some of the pitfalls encountered with NSPS. This system is still in development.

It is inte resting to note that som e military laboratories were exempt from NSPS and rem ained under experimental personnel systems designed for the laboratories. These systems have many of the features of NSPS but confer somewhat more authority to laboratory m anagement than did NSPS. Higher authorities control these laboratories in terms of how m any positions laboratories may fill and how much m oney the labora tories receive. Managers would prefer, given the program plans in the budget and laboratory funding, to be free to hire as m any people as they deem necessary to perform the mission and can affo rd. Again, this is may not be feasible in the Federal Government.

Other Ways to Acquire the Best and the B rightest. Sometimes the Governm ent's usual method of hiring em ployees is too cumbersome for a particular po sition, but other h iring processes are available. One is a no ncompetitive authority[11] for hiring STs, including those who are otherwise out of reach for the G overnment. These positions are generally the equivalent of the SES or general officer ra nk. The positions are not supposed to be m anagerial. Another mechanism under the Intergovernm ental Personnel Act (IPA) is an arrangem ent[12] whereby an employee of a non-Federal en tity may be detailed to a F ederal agency and have m ost of the rights and privileges—such as supervising others—of the agency. These em ployees, deemed IPAs, come for a time not to exceed 4 years—2 years plus an option for the second 2 years. The IPA may subsequently be hired for a regular government position, ex cepted or regular. IPA positions are used in many agencies that want to have a high rate of turnover, such as program managers at the National Science Foundation or DARPA, where fresh insights are needed for the work. IPAs can certainly be used at the Army laboratories.

Some laboratories participate in the National Research Council's (NRC) Research Associat e program for post-doctoral individu als (post-docs). (Other such post-doc programs include one conducted by the American Asso ciation of Engineering Education, which som e military laboratories use.) Intended for post-doc appoint ments for 1 or 2 years, the NRC vets the Associate and m atches him or her to requests for specific talent f rom the participa ting laboratories. Laboratory management can accept or reject a particu lar applicant. During the appointment, the post-doc can evaluate the la boratory for perm anent employment, and the laboratory can evaluate the post- doc for suitability for its program s. Conversion to perm anent employment is noncom petitive because the compet ition has been held at the NRC. These post-

---

[10] Public Law 111-84, Section 1113(d).
[11] See 5 U.S.C. 3104.
[12] Intergovernmental Personnel Act of 1970.

docs are among the very best candidates for the laboratories; experience has shown that about half of the post-docs are converted to regular permanent government appointments and represent a key cadre of researchers. Expanding this mechanism to cover more hires of PhDs would make sense.

Another option is for the laboratories to sponsor graduate work for promising young employees, including undergraduate interns. DOD operates a program that sponsors study for masters and doctorate degrees in science, mathematics, and research for transformation (SMART). Candidates may be for existing employees and for applicants from outside DOD. The recipients must commit to work in DOD programs after the degree program is completed.[13]

Visiting personnel from other government entities and from company and university laboratories can bring relevant expertise as well. Sometimes such visitors are part of a personnel exchange program in a formal research collaboration. Moving people is the best way to transfer information back and forth. The Army Research Laboratory's (ARL) Collaborative Technology Alliance (CTA) program includes a requirement for such personnel exchanges. These movements help keep personnel aware of the details of the technical work and the challenges that arise. Intimate collaboration in programs is a way of broadening the horizons of a laboratory and at the same time raising its visibility.

Leadership of the Laboratories. Effective leaders will ensure the laboratory's mission is clear; define a vision for success; and secure the necessary staff, funds, facilities, and equipment. They will oversee laboratory operations and represent them to the customers, Army overseers, Government, and public at large. Laboratory leaders should have an appropriate technical background and be able to strike a balance between defining the goals of the laboratory and micromanaging the details of the technical work. They should maintain awareness of progress by interacting with the staff in their workplaces but avoid telling them what to do. The selection of laboratory directors should be based on a mix of leadership skills and record of accomplishment in laboratory research. Aggressive National searches for candidates should be required for SES positions.

The Importance of Basic Research Programs. Ideally, a research laboratory should conduct a mixture of programs in basic research, applied research, and development. Basic research is important in providing a solid basis for the applied programs to come, attracting new PhDs and post-docs, and giving a laboratory the opportunity to expand into new areas. Basic work usually leads to associations with universities and thereby broadens the competence of the laboratory. We believe any Army laboratory should perform some basic research; the amount needed is debatable. We think a good balance among the S&T areas is something like the following:

- Basic research: 15 percent
- Applied research: 35 percent
- Advanced development: 50 percent.

These are not hard and fast numbers, but they do indicate what is needed and have withstood the test of time. Some Army laboratories meet this ratio; most do not.

---

[13] See "SMART Scholarship," American Society for Engineering Education at http:// smart.asee.org/

Assessment of Laboratory Qua lity. Assessments of the Arm y laboratories should be a continuing process. Effective assessments will reveal the kinds of im provements management should consider. One model of assessm ent (see ref. 8) considers three facets of a laboratory: quality, customer satisfaction, and relevance to Ar my needs. We consider quality assessm ent first.

Quality. We can assess technical quality as new work is proposed, as the research is in progress, and when the work is com pleted. Independent technical experts, usua lly external to th e laboratory, should review technical quality. E xperts designated by ARO always evaluate research proposals for external research gran ts. For proposals for work internal to the laboratories, the usual practice is for m anagement and perhaps senior scientists and engineers at the laboratory to perform the review. Larg e proposals must be defended in the budget formulation process. New starts related to speci fic military needs are developed in concert with the Army Training and Doctrine Command (TRADOC). The laborat ory director and designees, usually within the laboratory, review proposals to a laboratory director's discretionary fund for quality. Decisions are made on a number of criteria: the scientific basis, ability of the proposer to accomplish the task, relev ance as a high-p riority need for the warfighte r, and laborato ry resources to do the work. Often, the question is whet her it is better to stop something to address the new work. In m any environments, the budget and the staffing are zero-sum games; that is, there is usually no additional m oney for new thrusts, so some other program must give up some of its funding. Managem ent regularly assesses the quality of work in progre ss. In addition, an outside review is desirable to gain another perspective and an independent evaluation of the work quality, staff, environment, and infrastructu re. Our report on peer rev iew for lab oratories[14] summarizes the practices at a num ber of compar able laboratories. There is no o ne preferred model. Some laboratories cont ract with outside groups to c onduct the entire review process— from selecting the reviewers to m aking meeting arrangements and preparing and vetting the report. Others conduct their own review, sometimes using external experts. Some do not conduct an explicit peer review but rely on internal periodic briefings. Using an external contractor with high credibility, such as the NRC, tends to ensu re the report will be believable; internal reviews are less credible. Our analysis produced the following recommendations:

- The Army should establish a policy that the la boratories contract with outside groups to convene peer review panels and manage the review process.
- Panels of experts external to, and independent of, the laboratory should perform the review.
- Membership on the panels should be for defini te terms; conflicts of interest should be addressed.
- Reviews for a given component of a laboratory should be done every 2 or 3 years.
- Reviews should cover technical details at the project level.
- The panels should also assess the quality of the staff, the managem ent environment, the equipment, and the facilities.
- The panels should provide feedback to the la boratory staff and provide form al written reports.

---

[14] John W. Lyons and Richard Chait, *Strengthening Technical Peer Review at the Army S&T Laboratories,* Defense & Technology Paper 58 (Washington, DC: CTNSP, National Defense University, March 2009).

Under some circumstances, the chair of the exte rnal evaluation team may discuss the findings with senior leaders in the Arm y, the Office of the Secretary of Defense (OSD), and, perhaps, congressional committees. An addition al advantage of having ex ternal experts vis it the laboratory is that interactions with these knowle dgeable visitors broaden the staff perspective. We believe external peer review is very im portant in helping the laboratory strengthen its work and establish a reputation for its concern for technical quality.

The third kind of quality review is for publish ed work. For Army reports, each laboratory should have an internal review process. For papers submitted to archival journals, there should be two reviews—an internal review in the laboratory and an external re view by experts that is arranged by journal editors. Journal review usually leads to valuable sugges tions for the authors to add to or change portions of the paper. Publica tion in refereed journals is another wa y of calling attention to the laboratory and its staff and can aid in nominating staff m embers for various prestigious external awards and recognitions.

One of the authors (JWL) was at NIST for tw o decades and had the oppor tunity to observe NIST's evaluation program. For 50 years, NIST (formerly the National Bureau of Standards [NBS]) has contracted every year with NRC to perform an external peer r eview of NIST's laboratory programs. These review s are based on annual visits (recen tly changed to bienn ial) arranged by the NRC Board on Assessment of NB S; expert panels perform the reviews—one for each of the major organizational components. The results are published by the National Academies. Part of the review is to s ee how well the laboratory has addressed the recommendations in the previous report.

NIST was established as NBS by an Act of Congr ess in 1901. Part of the Act was the creation of a statutory visiting com mittee (VC) to review the status of NBS and report annually to the Secretary of Commerce and the Congress. The VC is concerned w ith broad issues of the laboratory as a whole. Until recently, at one of its meetings each year, the VC received an oral report from the chair of the NRC Board on Assessm ent as to the health of the tech nical work. The chair of the VC meets with the Secretary of Commerce annually. It had been the practice for the chair of the VC and the chair of the NRC board to testify together at hearings before the authorizing committees for NIST at both houses of Congress. Thus, NIST is assessed from the point of view of technical quality and policy concerns of th e stakeholders at C ommerce and Congress. The exposure also enhances the reputation of NIST.

The customer base for NIST is the worldwide sc ientific and engineering community, which is a broad and diffuse clientele. NIST has not attempted formal surveys of all these clients but rather relies on comments from the NRC reviews and the information gained from the many visitors to the laboratories and to management.

Customer Satisfaction. Customers want quality work that is timely and sharply focused on their requirements. Some are paying cust omers, and others receive the laboratories' products simply as part of the Arm y structure with no exchan ge of funds . The warfi ghters are the ultim ate customers of the Army laboratories. However, most of the laboratories' results are transmitted to other laboratories or to the acquisition comm unity's program managers. ARL, the corporate laboratory for the Army Materiel Comm and (AMC), sends m ost of its findings through the Research, Development, and Engineering Center s (RDEC), which in turn support the progra m managers and, sometimes, the contractors.

ARL surveys customer satisfaction annually. The technical directors of the RDECs meet at ARL every year to review the program and discuss i ssues they may have with the lab oratory. In addition, the RDECs m ake unofficial contracts with ARL covering some portion of the applied research (6.2). These contracts provide a good fo cus for the evaluation. In turn, the RDECs meet regularly with sponsoring progr am executive officers (PEO) and program managers to review progress against their goals. Relevance and timeliness are always a concern.

Stakeholders. Stakeholders are groups at the senior pol icy level with a vested inte rest in the laboratory. At NIST the stakehol ders are represented by the VC and the four congressional committees that handle authori zation, oversight, and budgeting. At ARL, there was, until very recently, a stakeholder group consisting of three-star generals at the deputy chief of staff level and chaired by the four-star commander of A MC. This group m et annually to discuss policy issues and the status of major program thrusts. (When ARL was placed under the new Research, Development, and Engin eering Command [RDECOM], the stakeholder group was disestablished.) The Army laboratory m anagement has no direct contact w ith top leadership at DOD or with the Office of Managem ent and Budget (OMB) and the Congress. Other stakeholders include TRADOC, its schools, and other commands based on their interest. The Combatant Commands (COCOM) are s takeholders as well. The AM C office occasionally schedules visits by laboratory staff to some of the COCOMs.

Adequacy of Funding for Basic Research and Recapitalizing Research Equipment. Two key attributes of a laboratory are the extent of its ba sic research program and the state of its research equipment and the faci lities. In one report, [15] we investigate the invest ment in these two areas. We believe a strong program in basic research is critical for an R&D lab oratory. The exact level required for effectiveness can be debated. W e have considered 15 percent of a laboratory's total funding to be a good target. The reasons for ha ving this am ount of ba sic work include (1) pushing out the frontiers of knowledge in topics of high interest to the Ar my, (2) providing a sound underpinning to the applied work, and (3) k eeping staff in touch with research leaders around the world through publication and scientif ic meetings. The presence of a strong basic research program facilitates the hiring of bright new PhDs because they will view the basic work as very similar to the research done for a PhD degree.

The three laboratories that are focused on research—ARL, Engineering Research and Development Center (ERDC), and Army Medical Research and Materiel Command (MRMC)— all have 15 percent or more of 6.1 funding. The RDECs have considerably less—some even have none. Some argue that these developm ent laboratories do not need 6.1 funding. But, for the reasons just presented, enough 6.1 funding to mount strong projects is necessary for the health and standing of these laboratories. We have suggested that the RDECs and related centers should have 6.1 funding at a level of 5 percent of in-hous e R&D funds, or preferably m ore. (One issue, however, is that there may be a level below which managers do not become skilled at managing 6.1 research. W e do not know wha t this level is .) We have also observed that a laboratory's culture has an im portant effect. If senior m anagers support basic re search, the staff will participate by submitting proposals. If management is indifferent, the staff will be as well.

---

[15] John W. Lyons and Richard Chait, *Assessing the Health of Army Laboratories, Funding for Basic Research and Laboratory Capital Equipment,* Defense & Technology Paper 72 (Washington, DC: CTNSP, National Defense University, September 2010).

Infrastructure Funding. We said ear lier that the quality of laboratory equipment and the facilities for performing R&D attract new staff. In one paper (see ref. 15), we assess the funding for capital labora tory equipment at the Arm y laboratories, as well as at other governm ent laboratories and one industrial la boratory. The basis for assess ment is shaky at best. The definition of capital eq uipment varies; the ro le of sponsors in buy ing capital equ ipment also varies and is not clearly reporte d; and the calculation m ay be based on core funding, total in-house funding, or total funding. These variations make assessments uncertain. However, the Army figures are internally c onsistent and are based on co re funding. The three corporat e laboratories—ARL, ERDC, and M RMC—report from 7 percent to 11 percent spending on capital equipment. These figures com pare favorably to th ose from NIST. NRL bases cap ital funding on total in-house funding, yielding about 2 percent. W hen computed based on core funding, this figure rises to near 9 percent, which is com parable to the Army corporate laboratories. The funding for the Arm y RDECs is substantially lower. However, som e of the RDECs receive equipment funded by their sponsors; we did not obtain these figures.

ARL and NRL estim ate that the recapitalization rate for equipm ent is about 60–70 years! The useful life of laboratory equipment is far less th an this, sug gesting that technical work is of ten done on equipment that is very old in origin but has been repeatedly upgraded. W e believe that up-to-date capital equipment is esse ntial for performing high-quality w ork and attracting high-quality employees. To evaluate how Arm y laboratories are faring in this rega rd, the lack of a common set of definitio ns must be addressed. We recommend that a n ew study be done on th e Army laboratories based on clear d efinitions and assumptions. We also urge that independent, external experts assess the status of laboratory equipment as part of regular peer reviews.

As for facilities funding, we earlier noted that the competition for military construction funding strongly tilts toward funds for soldier needs as opposed to laboratory need s. Were it not for the Base Realignment and Closure (BRAC) activities of the last 20 years, the funding situation for facilities would be even worse. However, we did not assess this.

Conclusions:
1. *The most important asset of a laboratory is its personnel.*

2. *A strong program of basic resea rch will ensure the laboratory is pushing the frontiers and exploring new areas. Basic research also is an attractor for hiring new staff from graduate school and bringing in post-doc associates.*

3. *Regular assessments of laboratory quality prov ide the basis for quality improvement. These assessments comprise peer reviews by externa l experts, customer reviews for timeliness and relevance, and stakeholder reviews for resource adequacy and program priority.*

CHAPTER V. PLANNING THE S&T PORTFOLIO

Although we did not set out to study the Army's planning of its S&T portfolio, our reports contain a number of studies that relate directly to planning. So me of the studies look back at what has been done, thereby revealing lessons for the future. In other studies, we have proposed new ways to plan and conduct the S&T program. We begin with what has been done, including interviews with soldiers, past attempts at forecasting tec hnology, and relations with other laboratories and industry. We then consider current efforts and present an analysis of the benefits of collaboration, sometimes on a global scale. Fina lly, we present the advantages of technology forecasting using concepts of converging S&T and urge that the Army use this approach.

A Look Back. In two papers,[16, 17] we discuss the interview and polling results from returning veterans, mostly from Iraq. We sought their comments on the efficacy of the tech nology they used in stabilization and reconstr uction activities, particularly re lated to gaps in the technology that they would like to see filled.

Soldiers agreed that there are significant problems in urban environments that current technology could not overcome. These problems include difficulty in tracking friendly forces; difficulty in handling vehicle traffic; poor non- line-of-sight communications; and lack of integrated planning for operations involving both military and non-military operations, such as convoy movements in cities. There are problems with ta ctical radio systems operated by our m ilitary and by our coalition and host-nation partners. Lack of language transl ation support continues to be a problem. Situational analysis in urban environments is dif ficult because GPS in com plex situations does not work very well. Soldiers cannot use their m ost potent weapons platform s, such as the Abrams tank, in urban combat for reasons of collateral damage.

Soldiers manning checkpoints need better ways to identif y individuals on the spot and m ore security when conducting security checks. Soldiers reported problems with conflicts in air space management, as well as poor availability and high cost of using unm anned aerial veh icles (UAV). Soldiers appreciate the new technologies regarding info rmation availability but are concerned with potential inform ation overload. They also com plained about the num ber and weight of b atteries they must carry. Power sy stems are needed for tem porary and perm anent bases.

Existing S&T program s are addressing som e of these problems; other program s need new or increased efforts. For exam ple, the Ar my has l ong conducted research on the battery problem and, lately, enhanced w ork on fuel cells. Resear ch on power and energy challenges is occurring for forward operating bases. [18] For our second paper (ref. 16), we developed a set of use cases

---

[16] Richard Chait, Albert Sciarretta, and Dennis Shorts, *Army Science and Technology Analysis for Stabilization and Reconstruction Operations,* Defense & Technology Paper 37 (Washington, DC: CTNSP, National Defense University, October, 2006)

[17] Richard Chait, Albert Sciarretta, John Lyons, Charles Barry, Dennis Shorts, Duncan Long, *A Further Look at Technologies and Capabilities for Stabilization and Reconstruction Operations,* Defense & Technology Paper 43 (Washington, DC: CTNSP, National Defense University, September 2007).

[18] John W. Lyons, Richard Chait, and James J. Valdes, *Assessing Army Power and Energy for the Warfighter,* Defense & Technology Paper 81, (Washington, DC: CTNSP, National Defense University, March 2011).

and presented them to warfighters at Fort Benning. [19] That paper presents a discussion of technologies for battle command followed by a discussion of the status of many technologies of general interest to the Arm y, including electronics, sensors, power sources for the individual soldier, basic research in adva nced materials, robotic system s, combat casualty care, and a synopsis of work being done by the Army on technology for battle command at the system level.

Figure 3 s ummarizes these d iscussions. It p resents the needed capabilities across sev   en technology areas. The objective of   this work was to help d   ecisionmakers in the Ar my S&T program gain insight into warfighter views and modify the S&T work accordingly.

	Command & Control	Comms	Computers	Intelligence	Surveillance	Recon	Force Protection	Unmanned Systems	Combat Care	Other
Electronics	●	●	●	●			O	●	●	O
Sensors	●		O	●	●	●	●	●	●	O
Materials			O				●	●		
Power Sources	O	O	O		O		O	●	O	
Robotics						O	O	●		O
Battle Command	●	●	●	●	●	●	●	●	●	●
Combat Medicine									●	

● = Strong Support   O = Moderate Support

Figure 3. Technologies—Capabilities Map in Summary (from ref. 17, p. 70)

Collaboration. Two papers (see ref. 20 and 22) discuss the advantages of collaboration between Army laboratories and efforts elsewhere in th   e Government and overseas. One discusses the interactions between the Ar my laboratories an d the research program s in industry conducted under the IR&D program. [20] The Congress established this program to stimulate R&D by DOD contractors with a focu s on future defense systems. Until 1992, Arm y experts evaluated each project in a program . In 1983, for example, there were about 30,000 evaluations on 10,000 projects. This m eant that the m ilitary had a ve ry detailed picture of th e industrial work being done. But the reviews were deem   ed too burde nsome, and the evalua tion requirement was dropped. Inevitably, DOD's knowledge of the industrial work has since declined; som   e laboratories have little or no idea of what is cu rrently going on in their areas. Other laboratories

---

[19] A use case presents a scenario with a set of explicit actions that must be accomplished to complete the mission. The capabilities required are then compared to available solutions – in our cases, technology gaps. At Benning there were lengthy discussions that gave a sense of the soldiers' priorities. An example was information overload; another was the need for persistent surveillance. The soldiers were worried about the ability of telecommunications to enable "a high degree of micromanagement from higher echelons." They also emphasized that proper training is necessary for soldiers to benefit from technology; they said in some cases training on new technologies being introduced in theater was too brief and superficial.

[20] John W. Lyons, Richard Chait, and Jordan Willcox, *Improving the Interface between Industry and rmy Science and Technology, Some Thoughts on the Army's Independent Research and Development Program,* Defense & Technology Paper 33 (Washington, DC: CTNSP, National Defense University, June 2009).

make an effort. One is the Army's Communications and Electronics RDEC (CERDEC); another is the Air F orce Research Laboratory (AFRL). Th ese have for mal programs to foster m eetings between government experts and their counterparts in the IR&D programs. In addition, the Navy has created a program not tied to IR &D but that develops the same understanding. The program also explicitly requires the Navy participants to brief th eir industry counterparts on Navy programs and needs.

We found (see ref. 20) that the Army is likely losing valuable information that could enhance Army S&T program s. We recommended that Ar my laboratories set up inform ation exchange programs with their industrial counterparts much as CERDEC has done. These program s need not be tied to the IR&D program.

In another paper, we discuss collaboration a nd the effects of globalization on the conduct of research.[22] A potential side benefit of collaboration and g lobalization is inc reased return on investment (ROI) on the military research done by the military partners. The paper begins with a discussion of a variety of forms of collaboration, ranging from informal individual pairings of researchers to form al long-term relationships with centers of excellence. The CTAs at ARL are true partnerships between cons ortia of companies and universities to carry out long-range basic research programs in specified areas. An exam ple area is that of micro-autonomous technology systems (see Figure 4). The C TAs are s tructured so ARL is actively involved in planning and execution. Also, staff rotation between ARL and th e consortia members is required. This rota tion facilitates early inf ormation exchange and, ultim ately, technology transfer to the Army. ARL also has an inte rnational version of the CTAs with the United Kingdom. These collaborations

Figure 4. A Microrobot Typical of the Focus of ARL's CTA in Micro-Autonomous Technology Systems[21]

differ from the m ore traditional centers of ex cellence, such as m ulti-disciplinary university research initiatives (MURI) and university affiliated research centers (UARC). The Army's three large centers of excellence— soldier nanotechnology, collabor ative biotechnology, and creative technology—do not involve Arm y technologists in day-to-day operations. The centers are m ore or less at arm 's length from the in-house research laboratories. Efforts should be m ade to tie these UARCs closely to the Army in-house S&T programs.

The international coll aboration agreement with the Unite d Kingdom may be the beginning of more global agreements that will ex tend the horizons of the Arm y laboratories. W e review the pros and cons of international work in our paper (see ref. 22). Of concern is always security . Basic research, usually considered open, is easiest to justify. Howe ver, there are som e countries

---

[21] Photo is from http://www.wired.com/dangerroom/2008/04/usually-our-dys/. Information on the CTA can be had from Dr. Joe Mait, ARL, Adelphi.
[22] John W. Lyons, *Army R&D Collaboration and the Role of Globalization in Research,* Defense & Technology Paper 51 (Washington, DC: CTNSP, National Defense University, July 2008).

where DOD has prohibited any collaborations in research. Berry and Loeb have urged that this restriction for China be removed for basic research.[23]

ROI in S&T. Stimulated by a U.K. Ministry of Defence (MOD) study on the ROI of its S&T work,[24] CTNSP was asked by its Army sponsor to look into the paper and offer a critique. The result was a set of three publications on the topic. There has been a great deal of effort by micro-economists to quantify the results of R&D. Doing so is relatively easy if one looks back far enough that a new product or process has matured and was based on identifiable technical contributions. Quantifying R&D results is very difficult to do in the near term, however, because financial impact usually comes years after the technical work is finished. The MOD study attempted to compare the ROI for battle tanks from different countries in terms of the countries' R&D investment relative to the capabilities of the resulting tanks. The study used panels of experts to judge the capabilities of the tanks in pairs. The judgment deemed whether each member of the pair was comparable to, better than, or poorer than the other. But this study did not differentiate between the quality of the differences; for example, the same result would have been obtained if a tank was a little better or a great deal better than the other in its pair. We decided to take another tack.

ROI analysis in the Government is not motivated by financial gains as much as by performance; there is no profit incentive. The Army can look at the value of benefits from its S&T program, if they can be quantified, compared to the S&T investment. Another approach is to estimate the costs incurred if no research is done; a third approach is to try to identify scientific measures of merit, which is useful in assessing basic research. These are discussed in the first of two reports[25] (see ref. 25 and 26). Cost is available in the records of the budget office, but benefits are not so easy to calculate. Sample benefits include risk reduction (improved survivability), capability enhancements (more lethality), and cost reduction. The value of the first two is difficult to quantify, whereas cost avoidance can be calculated. To estimate the potential of a given project proposal versus the likely benefits, one needs to estimate the probability of technical success by considering such factors as technical probability, probability of successful transition from the laboratory to the acquisition community, likelihood of fielding the results, and estimated length of time the result would ensure predominance over possible adversaries.

To test these estimations, we developed scenarios using microrobots for searching in-tact and demolished buildings and setting up perimeter defense for a special operations unit. Only the searching of buildings was proposed for the first test, which set the stage for the second paper. The second paper[26] presents the results of exercising the use cases at Fort Benning, Georgia. Since the technology had yet to be developed, a thought experiment (Gedanken) was conducted

---

[23] William Berry, Cheryl Loeb, *China's S&T Emergence, A Proposal for U.S. DOD – China Collaboration in Fundamental Research,* Defense & Technology Paper 47 (Washington, DC: CTNSP, National Defense University, March 2008).

[24] A.J. Middleton et al., *The Effect of Defence R&D on Military Equipment Quality,* Defence and Peace Economics, Vol.17, No. 2 (April 2006).

[25] Albert Sciarretta, Richard Chait, Joseph Mait, and Jordan Willcox, *A Methodology for Assessing the Military Benefits of Science and Technology Investments,* Defense & Technology Paper 55 (Washington, DC: CTNSP, National Defense University, September 2008)

[26] Albert Sciarretta, Joseph Mait, Richard Chait, Elizabeth Redden, and Jordan Willcox, *Assessing Military Benefits of S&T Investments in Micro Autonomous Systems Utilizing a Gedanken Experiment,* Defense & Technology Paper 80 (Washington, DC: CTNSP, National Defense University, January 2011).

with a team of technical personnel from ARL, members of the ARL CTA on micro-autonomous systems, and soldiers with experience in small unit combat operations. The experiment involved the three scenarios posed in the first report (see ref. 25): clearing a building, defending the building once cleared, and dealing with an explosion in an urban setting with snipers present and possible insurgents in the local population. The S&T participants learned many things from the warfighters that changed the S&T perspective. The interactions benefited the soldiers and the researchers in documented ways. There is no question that the joint exercise between research staff and experienced combat soldiers sharpened the focus of the research program and guided future work. The Gedanken Experiment with the combination of soldiers and technologists has been judged a success; more such experiments are being contemplated for other areas. For example, one on network science in small unit operations has just been completed at Fort Benning. Participants in these experiments have been enthusiastic and supported more such exercises. Appendix D has a summary of the latest Gedanken Experiment. Because of technologists' close collaboration with warfighters, Gedanken Experiments are likely to find a permanent role in planning the S&T portfolio.

Technology Forecasting. The Army conducted a full-scale effort in 1992 known as STAR 21 to forecast developments in S&T.[27] We assessed that report through interviews with SMEs to see how well the predictions held true and documented our results in a paper. [28] As might be expected, some predictions were right on target, and some underestimated progress (e.g., in optoelectronics and photonics and manufacturing at the nano-scale). Some projections were overly optimistic; some were outright wrong. One or two were, in our estimation, way off the mark (e.g., the use of ground- or space-based free electron lasers for destroying missiles). STAR 21 missed some important areas: the evolution of computing and the Internet, the wireless revolution, information security, information overload, and micro-UAVs. Nonetheless, it was a valuable exercise in terms of educating the nontechnical people overseeing and funding the S&T enterprise. STAR 21 also influenced the thinking of the S&T managers. We recommended another round of forecasting occur. In our paper (see ref. 28), we recommend that such forecasting be done at least every decade and that it be broken up into sections conducted in a rolling manner so all sections are not forecast at once. We also recommend that areas with similarities be grouped together; the same was suggested for basic sciences with their related technologies. We believe that more participation by Army SMEs can bring to the proceedings a better understanding of Army challenges. The Army at the level of DAS(RT) decided to pursue these recommendations, so we undertook studies on current methods of forecasting and made recommendations in two succeeding documents.[29] We urged that the next studies be done by the three services working together under the Assistant Secretary of Defense for Research and Engineering and proposed that a relatively new concept based on the convergence of individual

---

[27] National Research Council, *Star 21 – Strategic Technologies for the Army of the Twenty-First Century,* (Washington, DC: National Academies Press, 1992).

[28] John Lyons, Richard Chait, and Jordan Willcox, *An Assessment of the Science and Technology Predictions in the Army's STAR 21 Report,* Defense & Technology Paper 50, (Washington, DC: CTNSP, National Defense University, July 2008).

[29] John W. Lyons, Richard Chait, and James J. Valdes, *Forecasting Science and Technology for the Department of Defense,* Defense & Technology Paper 71, (Washington, DC: CTNSP, National Defense University, December 2009); John Lyons, Richard Chait, and Simone Erchov, editors, *Improving the Army's Next Effort in Technology Forecasting,* Defense & Technology Paper 73, (Washington, DC: CTNSP, National Defense University, September 2010).

areas of science or technology could be more powerful than considering areas in isolation. As an example, we presented a subset of one such study [30] (see Appendix A) that consisted of forecasting future convergences in four areas of biotechnology— convergences that could lead to new, perhaps unanticipated, products and capabilities.

Unfortunately, the three services could not agree on launching a new tri-service series of forecasts (the Navy and the Air Force had also published forecasts in the 1990s similar to STAR 21), so the Army decided to proceed with its own forecasting. The second paper in ref. 29 details how such studies might be done. The concept involves selecting several separate sciences or technologies closely enough related to provide possible convergences and studying them together. We decided to couple the forecasts to desired outcomes as given by TRADOC. With such outcomes or capabilities in mind, the forecasts would look for convergences that would enable the realization of the desired capabilities.

Experts in four fields contributed essays from their fields. The paper contains two examples of physical sciences and engineering: one in the human dimension and one in virtual presence, enhanced sensing, and augmented autonomy; and two examples of basic science: one on mechanochemical transduction and one on quantum information science. In each of the four cases, in addition to presenting the technical details of the several related disciplines suggested for forecasting, we included a chart showing the possible evolution from the separate areas to convergence to products or capabilities that would realize the desired outcomes. In the case of human dimension, six areas—physics, materials science, mathematics, neuroscience, biochemistry, and pharmacology—are predicted to evolve through two intermediate steps to deal with preventing or treating post-traumatic stress disorder. The paper concludes with a suggested methodology for conducting such studies and suggested ways of presenting the results. In it, we urge the Army to undertake a test case that could provide a template for further studies.

ASAALT accepted our suggestion for a pilot study, and ARO agreed to manage a study on mechanochemical transduction—a new area of interest in the chemical and materials sciences. ARO very recently conducted the workshop, where academics from departments of chemistry and materials science and engineering participated along with people from ARO, CTNSP, ASAALT, ARL, Tank and Automotive Command, and TRADOC. The objective of the workshop was to demonstrate the utility of the approach and to provide a template for future technology forecasts. The academics presented 5-minute talks on their research related to the topic; the group then split into two breakout locations to develop forecasts in their specialties and to look for potential convergences to provide new and perhaps unexpected opportunities. Participants were instructed to keep in mind the Army's priorities for new capabilities. For more on the workshop, see Appendix C.

One area that would benefit from a convergence forecast is power and energy for the warfighter. In the paper in ref. 18, we urge that a convergence forecast be conducted in this area. We recommend that a more holistic approach be taken, especially for isolated forward combat operations posts. The idea of holism, taking into accounts all aspects of a situation or system, is consistent with the idea of convergence, which attempts to add dimensions to the forecasts. In addition to suggesting a forecast, the power and energy report notes serious problems in

---

[30] Unpublished information from James J, Valdes, CTNSP, National Defense University, Washington, DC.

calculating the fully burdened costs of energy on the battlefield and urges that the military apply more sophisticated expertise in doing such analyses.

It is difficult to launch new approaches to power and energy on the battlefield without knowing the economics of the current approaches. In ref. 18, we also propose that the Army mount more basic research in this area and tie the sponsored academic research at ARO more closely to the applied work at the Army laboratories. Finally, it is worth noting again that the Army does not seem to be aware of several relevant studies by NRC. For example, one Army white paper[32] does not cite any of the several recent comprehensive studies by NRC.[33]

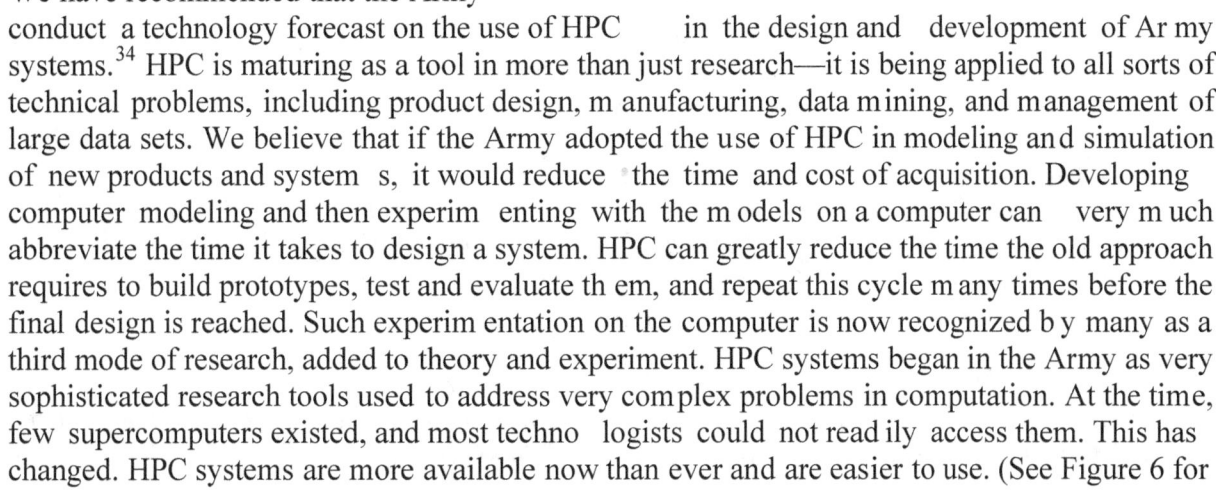

Figure 5. Chart From a Study by ARL[31] (AOE [Army of Excellence] represents the Army logistics as of 1998; Army XXI represents the planned digitized Army as of 1998; AAN [Army After Next] is a projection into the next century)

The Use of High-Performance Computing (HPC) to Strengthen the Development of Army Systems. We have recommended that the Army conduct a technology forecast on the use of HPC in the design and development of Army systems.[34] HPC is maturing as a tool in more than just research—it is being applied to all sorts of technical problems, including product design, manufacturing, data mining, and management of large data sets. We believe that if the Army adopted the use of HPC in modeling and simulation of new products and systems, it would reduce the time and cost of acquisition. Developing computer modeling and then experimenting with the models on a computer can very much abbreviate the time it takes to design a system. HPC can greatly reduce the time the old approach requires to build prototypes, test and evaluate them, and repeat this cycle many times before the final design is reached. Such experimentation on the computer is now recognized by many as a third mode of research, added to theory and experiment. HPC systems began in the Army as very sophisticated research tools used to address very complex problems in computation. At the time, few supercomputers existed, and most technologists could not readily access them. This has changed. HPC systems are more available now than ever and are easier to use. (See Figure 6 for

---

[31] *The Impact of Fuel Efficiency on the Army After Next,* ARL briefing FEAAN Ia, October 1998 available from the office of the director, Sensors and Electron Devices, ARL, Adelphi, Maryland.

[32] Col. R. C. Effinger, *Warfighter Challenges/Warfighter Outcomes,* Presented at the Technology Planning Conference, ARL at Adelphi, Maryland, May 2010; Army Capabilities Integration Center; Research, Development, and Engineering Command; and the Deputy Chief of Staff, G-4, *Power and Energy White Paper,* published by TRADOC, Fort Monroe, VA, April 2010.

[33] See John W. Lyons, *Assessing and Predicting for Army Science and Technology,* the second paper in Defense & Technology Paper 12 (Washington, DC: CTNSP, National Defense University, March 2005, p. 44).

[34] John W. Lyons, Richard Chait, and Charles Nietubucz, *The Use of High Performance Computing to Strengthen the Development of Army systems,* Defense & Technology Paper 43 (Washington, DC: CTNSP, National Defense University, November 2011).

one example of an HP C.) Because of the adv ent of the Internet and its fiber op tics links, an engineer can sit at his or her desk and work w ith an HPC system even though it is located in another state or overseas. The software is improving and will be more scalable in the future.

Army systems are designed cooperatively through Ar my laboratories and centers, PEOs and their program managers, and contractors who are bidding on the production work. When this work is done on HPC system s shared in real tim e over the network, the overall tim e and cost of the developm ent cycle is reduced significantly. In one case study reported by industry, this savings was a factor of three in the overall developm ent time.[35] Adopting this approach to developing Army systems or modifying existing systems can produce savings in both tim e and money in the acquisition process.

Figure 6. The IBM Blue Gene P Supercomputer Runs at Petaflop Speeds

We believe Arm y S&T program planning will benefit from adopting th e above suggestions. Some of these require simple instructions through the management chain from ASAALT. Some may require policy m emoranda from ASAALT in coordination with affected commands. Som e may require additional resources and funding r eallocations. But all are straightforward actions based on experiences in other respected R&D programs.

Conclusions:

1. *Close interactions with soldiers improve the usefulness of Army S&T. Understanding deficiencies in technology on the battlefield by talking to returning soldiers sharpens the focus of research programs. The value of use cases and thought (Gedanken) experiments have been shown to be helpful in analyzing and planning.*

2. *Formal collaborations with academe, industr y, and other public la boratories are used increasingly in Army research. The Army does not have a good understanding of the industry work done under the IR&D program. This is a missing value.*

3. *Technology forecasting is of value to managers of S&T; the last major forecast for the Army was done in the early 1990s. A new approach invo lving the study of potential convergences of various S&T disciplines has been adapted for Army resear ch. A pilot study has been conducted and should lead to a template for broad forecasting in the future.*

4. *The Army should increase the use of HPC, especially in system design and development.*

---

[35] Loren Miller in *Goodyear Puts the Rubber to the Road with High Performance Computing,* case study available from the Council on Competitiveness, Washington, DC. 2010; for more details see Loren K. Miller, *Simulation-Based Engineering for Industrial Competitive Advantage,* Computing in Science & Engineering, May/June, 2010.

CHAPTER VI. MANAGING THE LABORATORIES— INDIVIDUALLY AND COLLECTIVELY

CTNSP has recently given thought to the ph ilosophy of, and guidelines for, m anaging R&D in DOD laboratories. One study began after the Departm ent of Ho meland Security (DHS) asked CTNSP staff with relevant DOD ex perience for advi ce on a series of issues that had arisen in DHS. This request stimulated th inking about aspects of laborat ory management that had not previously occurred to m embers of the CTNSP group. The quest ions touched on areas we, as long-time R&D managers, had taken as givens. The results of this thinking are useful for this paper as well. A second study consisted of inte rviews with three reti red directors of DOD laboratories—one each from the Navy, Ar my, and Air Force—perf ormed to obtain person al opinions on and practices for dealing with issues these directors deemed important. Also, one of the authors of the present paper (JWL) wrote a memoir on his experiences managing laboratories in industry and in two governm ent agencies. Th at paper considers a full set of m anagement challenges and discusses how he handled them.

DHS, which was formed after 9/11, needed to pull together a group of laboratories that DHS had inherited from the vario us agencies from which its function s were drawn. These ag encies had diverse charters and experiences and were resp onding to new challenges arising from 9/11. DHS sent us a list of questions focused on managing its R&D portfo lio using risk m anagement. By this, DHS meant balancing potential payoff fr om a project with the p otential for failure caused by any of a num ber of negative factors or risks. In our first resulting report, [36] we list 11 risk factors taken fro m our experiences in DOD, such as the lack of su fficient expertise in the technical staff and failure to secure adequate f unding. We point to the need for relevance to the mission and the lack of support from the custom er. We also consider the difficulty of the technical challenge—sometimes a project is just too hard or would take much too long.

Next, the report lists 10 com ponents of i mpact when a project succeeds. These include value to the mission, wide applicability of the results beyond the use of the sponsor, the addition of valuable information to the knowledge base in the com munity at large, and avoidance of technical surprise by being first with new results. For Federal labor atories, adding to the profits of a sponsoring organ ization is not a factor. By applying arbitrary numbers to each of the risks and impacts, we suggest that figures of m erit can be given to a technical proposal. W e also offer a flowchart for decisionmaking based on these factors.

In a follow-up paper, [37] we cover issues in m anagement beyond project selection. Topics include workforce management, relations with outs ide SMEs, relations with the u ltimate user community, balance in the program —especially long-term fundamental work (basic research) to keep the laboratory at the cutting edge of S&T, ability to accept sponsored funding, nonmanagerial career paths for top technical staff, excellent equipment and f acilities, ability to publish results and attend professional m eetings, and independent outside peer review of the

---

[36] Samuel Musa, William Berry, Richard Chait, John Lyons, and Vincent Russo, *Risk-Informed Decision making for Science and Technology,* Defense & Technology Paper 76 (Washington, DC: CTNSP, National Defense University, September 2010).

[37] Samuel Musa, Richard Chait, Vincent Russo, and Donna Back, *Strengthening Government Laboratory Science and Technology Programs: Some Thoughts for the Department of Homeland Security,* Defense & Technology Paper 83 (Washington, DC: CTNSP, National Defense University, July 2011).

laboratory's ongoing work. The re port concludes with discussi on of leadership and a differentiation between a m anager and a leader. Ideally, a labor atory director should have elements of both. Technical managers normally consider all of these factors when planning their research portfolios.

Interviews With Former Executiv es of DOD Laboratories. We gained im portant insights from a serie s of interviews with retired execu tives of the th ree service corporate laboratories: NRL, ARL, and AFRL.[38] Comments from a former senior official at the o ffice of the Assistant Secretary of Defense for Research and Engineering and from a book written by a former director of that offi ce elaborated on the executiv es' views.[39] All of the inter viewees stressed the importance of having a robust basic research pr ogram for the reasons already discussed. All believe personnel are the key to la boratory success and that, ther efore, managers must address the factors that are im portant to research professionals, including a ch allenging mission; first-class equipment and facilities; stimulating colleagues; supportiv e management; and freedom t o publish, obtain patents, and travel to technical meetings and laboratories with similar interests.

One of the interviewees (JWL) recently wrote a m emoir of his 50+ years in R&D. [40] In it, he discusses in detail his experiences as an R&D manager at three different laboratories—one in the chemical industry and two in the Federal Governm ent. The m emoir begins with a general discussion of research in scie nce and engineering intended as a p rimer on working in a laboratory. The discussion is aimed at young people studying for a career and m anagers responsible for R&D but who lack direct experience (e.g., senior ex ecutives in companies or the Government who have a research function under them and who m ake decisions about budgets and facilities). The rest of the memoir deals directly with the subjects discussed in this chapter.

Managing the Collection of Army Laboratories. The foregoing has been focused on managing an individual laboratory and all the factors a laboratory m anager should keep in mind to ensure the laboratory rem ains strong and effective. A n additional considera tion when managing or overseeing the set of Ar my laboratories includes placem ent of the labo ratories in hos t organizations—the level and additional lines of responsibility and oversight that may be needed. The Army has three central laboratories: AMC's ARL, the Corps of Engineers' ERDC, and the Army Medical Comm and's MRMC. Their reporti ng level varies. AMC's corporate laboratory, ARL, reports to the AMC RDECOM, which in tu rn reports to the Comm anding General of AMC. There is no f ormal relationship between ARL (with its ARO) and the Of fice of the ASAALT. At the Corps of Engineers, th e Director of Research and Development at Corps Headquarters also serves as the Di rector of the ERDC. In both pos itions, he reports to the Chief of Engineers. The Ar my MRMC reports to the Army Medical Departm ent, headed by the Surgeon General. MRMC has within it five major research areas that are coordinated and report to the Comm and's office. Thus, one of the resear ch functions (ERDC) re ports directly to the Commanding General of the major command, and the others report through intermediates levels. NRL reports to the Chief of Naval Research and through him to both the Chief of Naval

---

[38] Richard Chait, *Perspectives from Former Executives of the DOD Corporate Research Laboratories,* Defense & Technology Paper 59 (Washington, DC: CTNSP, National Defense University March 2009).

[39] Hans Mark and Arnold Levine, *The Management of Research Institutions* (Washington, DC: National Aeronautics and Space Administration, 1994).

[40] John W. Lyons, *Reflections on Fifty Years in Research and Technology,* Defense & Technology Paper XX (Washington, DC: CTNSP, National Defense University, in press, 2012).

Operations and the Assistant Secretary of the Navy for Research, Development, and Acquisition. AFRL reports to the Commanding General of the Air Force Materiel Command.

At NIST, the directo r also serves as the Under Secretary of Comm erce for Standards and Technology. The director is nominated by the President and confirmed by the Senate. He reports directly to the Secre tary of Commerce and in teracts closely with the v arious committees of the Congress. At NIH, the director is also nom inated by the President and confirm ed by the Senate. He reports to the Secretary of Health and Huma n Services. These officials work directly with OMB and with the res ponsible committees in the Congress. This m akes realization—by direct communication and form al testimony—of their or ganizations' technical needs possible. In contrast, directors of Army labor atories rarely have exposure to the people who control their budget destiny.

In the private sector, the central research d epartment of a corporation usua lly reports to a senior vice president who may also be a m ember of the board of directors. This was true of the Bell Laboratories, an independent subsidiary that reported to the board of AT&T. At IBM, the central laboratory reports to a senior vice president. Titles a t various companies include Chief Technology Officer, Vice President for Science, and Vice President for Research—all of who m are at the top of the reporting chain. These arrangements are typically in firms that depend on an aggressive R&D program to rem ain competitive. The Army structur e appears to keep the laboratories, especially in AMC, at arm's length from top decisionmakers.

Some Studies of Government Laboratories. There have been m any studies of the Government's laboratories, m any of which have been in DOD. So me studies consider the placement of the laboratories in or out of th e Government. Some consider breaking up the assignments to the laboratories and hiving them off to va rious recipients. Others consider the nature of the laboratories' work; for exam ple, basic research and creativity. One report considers the National laboratories of DOE and m akes sweeping recomme ndations for change in the management paradigm (see the next section).

In a series of reports, the Defense Science Board (DSB) considers the defense laboratories.[41] The DSB suggests that the laboratorie s make massive use of personnel under the IPA in lieu of civil servants. DSB also suggests abolishing the labo ratories, transferring the S&T work (6.1, 6.2, and 6.3) to universities, and transferring 6.4 and ab ove to the acquisition co mmunity. Coffey et al. have analyzed the positioning options. [42] They list four possibilities: governm ent owned-contractor operated (G OCO), federally funded research and developm ent centers (FFRDC), public-private partnerships, and governm ent-owned corporations. Coffey et al. favor a government-owned corporation similar to the Te nnessee Valley Authority or the Saint Lawrence Seaway Development Corporation. Perhaps a more useful m odel would be a state-owned university. A university is more like a large Federal laboratory than the examples above.

---

[41] *Report on the 1987 Summer Study on Technology Base Management,*(Washington, DC: Department of Defense, December, 1987); *Defense Science Board Task Force on Defense Laboratory Management, Interim Report* (Washington, DC: Department of Defense, April 1994); *Defense Science Board Summer Study on Defense Science and Technology* (Washington, DC: Department of Defense, amy, 2002).

[42] T. Coffey, K. Lackie, and M. Marshall, *Alternative Governance: Tool for Military Laboratory Reform,* Defense Horizons 34 (Washington, DC: CTNSP, National Defense University, November 2003).

In 1993 the NRC's Board on Ar my Science and Technology (BAST) studied, at the Ar my's request, four new organizational and management options that m ight be m ost suitable for the newly established ARL.[43] They found that converting ARL to a GOCO would entail a sizeabl e up-front cost and the resulting en tity would be at arm 's length f rom the custom er—the warfighter. The BAST considered an enhanced ARL option and a NIST -like option. The BAST focused on what it termed a multi-center option. This option would contract out up to roughly 70 percent of the technical work but keep a sm all staff of program managers and some technical work. The model is a com bination of the en hanced ARL concept an d the GOCO idea. The enhanced ARL approach would maintain ARL as an Army laboratory but would grant additional authorities to ARL management, including expanded powers in personnel, procurem ent, and the like—some of which were already part of refo rms mounted in the 1980s. The m ultiple centers would be under contract to ARL and would con duct research in areas w here the centers would have the best expertise to do th e job. These en hancements continue. The contracting to several centers has begun under the CTAs. The CTAs have succeeded and are growing in num ber. Finally, the BAST recommended that ARL report directly to ASAALT, thereby making clear the special status of ARL. Such a repo rting channel would make ARL si milar to NRL and also to NIST. Many of these ideas have b een implemented; others are s till—almost 20 years late r— being considered. Two recent studies have proposed the idea of raising the repo rting level for ARL.[44]

In these not-yet-released studies, former senior leaders have looked at the report ing levels in the acquisition system and have recommended changes, primarily to tighten the reporting chain and focus more sharply on the m ission. ARL was orig inally established as a subcomma nd of AMC reporting directly to the Commanding General. However, in a subsequent AMC reorganization, RDECOM was established under AMC and t he laboratories were placed under it. Thus, ARL reports through two layers to reach the Army Chief and has no f ormal ties to ASAALT. Interactions with higher levels in the Pentagon, OMB, and the Congress are strictly limited.

In ref. 44, two points are m ade regarding Army laboratories. One is to disestablish RDECOM at AMC and assign the R DECs to their form er homes in the Life Cycle Management Commands. Presumably, this suggestion w ould leave a question as to wh ere ARL should be located. The second option worthy of consideration is to leave as is the organizationa l structure but create formal dotted-line relationships between the laboratories and ASAALT . In this option, the directors would be rated in AMC, but the secon d-level raters would be in ASAALT. In m atters such as major appo intments in the labor atories and cr itical personnel actions, including prioritizing personnel headcount in budget deci sions, ASAALT would have to concur. Yet another option would be to transfer ARL/ARO to ASAALT but leave the RDECs in RDECOM.

The National Laboratories of DOE: The Galvin Report. In 1995 a task force of the Secretary of Energy Advisory Board reporte d on its study of the future of the National laboratories. T he report (commonly called the Galvin Report after the Task F orce Chairman Robert Galvin, CEO of Motorola) begins by stating that "government ownership and operation of these laboratories

---

[43] *The Army Research Laboratory: Alternative Organizational and Management Options* (Washington, DC: Board on Army Science and Technology, National Research Council, 1994).
[44] Unpublished; under study by the Army. For some insight into one of these see "*Study Paints Bleak Picture of Billions Sunk into Incomplete Army Programs,*" Inside the Army, February 14, 2011.

does not work well." [45] The history of the 10 laboratories studied goes back to the scientific and development efforts in World War II when the Federal Government called on industry to help with the development and production of radar, the atomic bomb, proximity fuses, and the like. After the war, some of those laboratory sites were continued under a contract management arrangement with industry or universities; these were termed GOCO facilities. The original concept was to gain the flexibility and quick-response capabilities of the private sector while doing the Government's technical work. And, initially, this was the case. But as time went on, laboratory management by the Government became tighter and more bureaucratic to the point where many of its advantages were lost. Hence, the Galvin Task Force's assignment to see what could be done to loosen Federal oversight.

The Galvin Task Force was convinced that something very different needed to be done. It recommended an experiment in "corporatizing" the laboratories. In this model, the laboratories would be outside government control and funded by large blocks from the Congress. The weapons laboratories would likely be excluded. Each of the laboratories (or groups of laboratories) would be established in corporations managed by Boards of Trustees. The role of DOE would be to serve as the principal customer of the laboratories. The trustees would decide how to allocate funding across the laboratories and topical areas. The management style would not be according to the "Federal norm." Certain essential governmental services would be treated as exceptions. The corporatized laboratories would be able to serve customers other than DOE.

This proposal is similar to the public corporation discussed by Coffey et al. To our knowledge, the Galvin Report's recommendations on organization and management have not been implemented, but DOE has attempted a number of reforms. The first Packard report in 1983 proposed some of these reforms,[46] including setting up director's discretionary funds for new initiatives, and DOE did so.

Conclusions:
1. *Drawing on the experiences of several CTNSP senior staff and outside experts, a set of factors necessary for successful S&T programs emerged, including having an outstanding staff, a strong basic research program, and good laboratory equipment and facilities. Motivation for technical staff consists of a rewarding laboratory mission; accomplished colleagues; supportive management; and freedom to publish, patent, and travel to technical meetings and other laboratories. Salary is not a leading factor as long as it is adequate and fairly administered.*

2. *The reporting level of the laboratories is important as an expression of the priority given to them in budgeting the funding and staffing levels. The question of the laboratories' organization within AMC has often been raised. Options have ranged from no change to restructuring the laboratory reporting relationships.*

---

[45] *Report of the Task Force on Alternative Futures for the Department of Energy National Laboratories,* prepared by the Secretary of Energy Advisory Board, February 1995.
[46] *Report of the White House Science Council Federal Laboratory Review Panel* (Washington, DC: The White House, May 1983).

CHAPTER VII. SUMMARIZING THE CONCLUSIONS— RECOMMENDATIONS

In this paper, we have discussed important factors that can improve Army S&T.

In Chapter II, we stud y the role of the Army laboratories in the fielding of m ajor Army warfighting platforms. *In Project Hindsight Revisited, we learned the importance of close working relationships between i ndustry scientists and enginee rs and the Army laboratories and acquisition specialists. Success in fielding major Army platforms is attribu table to this teamwork; lack of it would have led to failure.*
Recommendation: *Continue to emphasize the importanc e of maintaining close working relationships with the technical staffs of Army contractors.*

In Chapter III, we consider the reas ons for the continuing criticisms of the laboratories and the suggestions that they be privatiz ed. We conclude that the cause of these critiques is in part the inability of the Army S&T program to publicize the contributions of the laboratories and that this has led to a poorer reputation than is warranted. *Enhancing the reputation of the laboratorie s requires that the quality of the work be maintained at the highest level.*
Recommendation: *Implement an aggressive campaign to publicize the technical contributions of the laboratories to the senior leadership of the Army and to the general public.*

We conclude in Chapter IV *that the most important asset of a laboratory is its personnel.*
Recommendation: *Pay constant attention to all aspects of Army l aboratory management that affect the ability of laboratory to hire and retain top-flight technical personnel. Take particular care to ensure the "stars" are handled properly. This includes paying attention to how the Ar my STs are managed.*

We further conclude that *a strong program of basic research should ensure the laboratory is pushing the frontiers and exploring new areas. Basic research also is an attractor for hiring new staff from graduate school and acquiring post-doc associates.*
Recommendation: *Ensure each laboratory has a significant level of funding for basic research.*

We are convinced that *regular assessments for laboratory quality provide the basis for quality improvement. These a ssessments comprise peer reviews by external expe rts, customer reviews for timeliness and relevance, and stakeholde r reviews for resour ce adequacy and program priority.*
Recommendation: *ASAALT should require, and each Army laboratory should arrange for, regular quality assessm ents by independent, exte rnal SMEs. Customer eva luations should also be required. Stakeholders should be alert for possible problems that may arise.*

In Chapter V, we demonstra te that *close interactions with soldiers improve the u sefulness of Army S&T. Understanding deficiencies in technology on the battlefield by talking to returning soldiers sharpens the focus of research programs. The value of use cases and thought (Gedanken) experiments has been shown to be helpful in analyzing and planning.*

Recommendation: *Continue to increase interactions with experienced soldiers in defining and shaping research programs. The successes with use cases and thought experiments at Fort Benning justify more such cooperation with soldiers.*

We further conclude that *formal collaborations with academe, industry, and other public laboratories are used increasingly in Army research. The Army does not have a good understanding of the industry work done under the IR&D program. This is a missing value.*
Recommendation: *Undertake more formal collaborations with industry's IR&D programs (pertains to all Army laboratories).*

The Army has developed useful collaborations with consortia of companies and universities under formal arrangements. At ARL these are called CTAs. They have been successful as evidenced by the extension of some and the creation of a number of new ones over the last dozen years.
Recommendation: *Continue to use this mechanism for formal collaborations.*

Technology forecasting is of value to managers of S&T; the last major forecast for the Army was done in the early 1990s. A new approach involving the study of potential convergences of various S&T disciplines has been adapted to Army research. A pilot study has been conducted and should lead to a template for broad forecasting in the future.
Recommendation: *ASAALT should conduct periodic technology forecasting. This should be based on the concept of convergences among fields of S&T. Frequency should be such that every priority area is covered every decade. Studies should be oriented to needed capabilities defined by the warfighters.*

In response to recent suggestions that the Army must rethink its acquisition processes to markedly reduce cost and shorten the time to complete a program, we studied the recent beginnings of a trend to use HPC to achieve these effects. This signifies the movement of HPC from the research environment to the commercialization of new technology.
Recommendation: *Expand the use of HPC in the design and manufacture of Army systems.*

In Chapter VI, we conclude that *drawing on the experiences of several CTNSP senior staff and outside experts, there emerged a set of factors necessary for successful S&T programs. These include having an outstanding staff, a strong basic research program, and good laboratory equipment and facilities. Motivation for the technical staff consists of a rewarding laboratory mission; accomplished colleagues; supportive management; and freedom to publish, patent, and travel to technical meetings and other laboratories. Salary is not a leading factor as long as it is adequate and fairly administered.*
Recommendation: *ASAALT should make more use of the experiences of retired managers of R&D, especially from other parts of DOD and from other Federal laboratories.*

The question has often been raised of the organization of the laboratories within AMC. Options range from no change to restructuring the laboratory reporting relationships and moving the corporate laboratory either to AMC headquarters or up to ASAALT.
Recommendation: *Seriously consider reorganizing the AMC laboratories.*

ASAALT should be actively engaged in solving a ll major issues relati ng to the laboratories . ASAALT should have shared autho rity in such matters as hi ring senior managers and setting management priorities in relation to downsizing.

Recommendation: *Manage the Army laboratories as th e important component of acquisition that they are. Emphasize the reporting rela tionships and the role of ASAALT in develop ing policy affecting the laboratories.*

APPENDIX A. EXAMPLES OF CONVERGENCE STUDIES (PROPOSED)

(from ref. 25, Defense & Technology Paper 73)

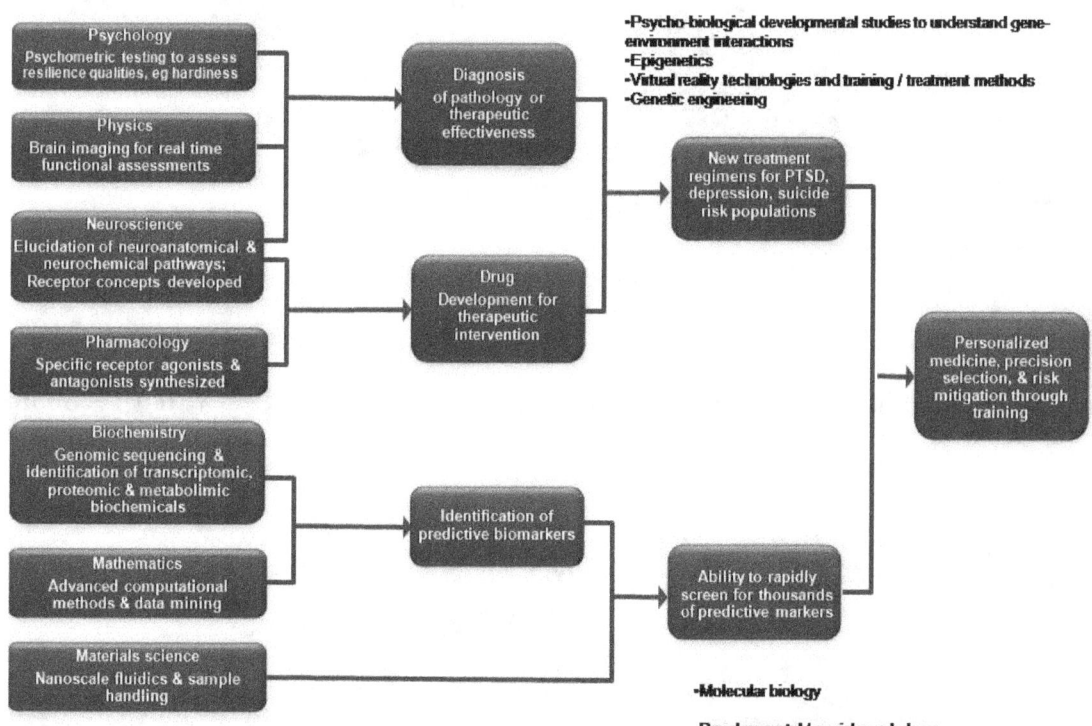

Figure 7. Convergence of S&T: Biobehavioral Resilience

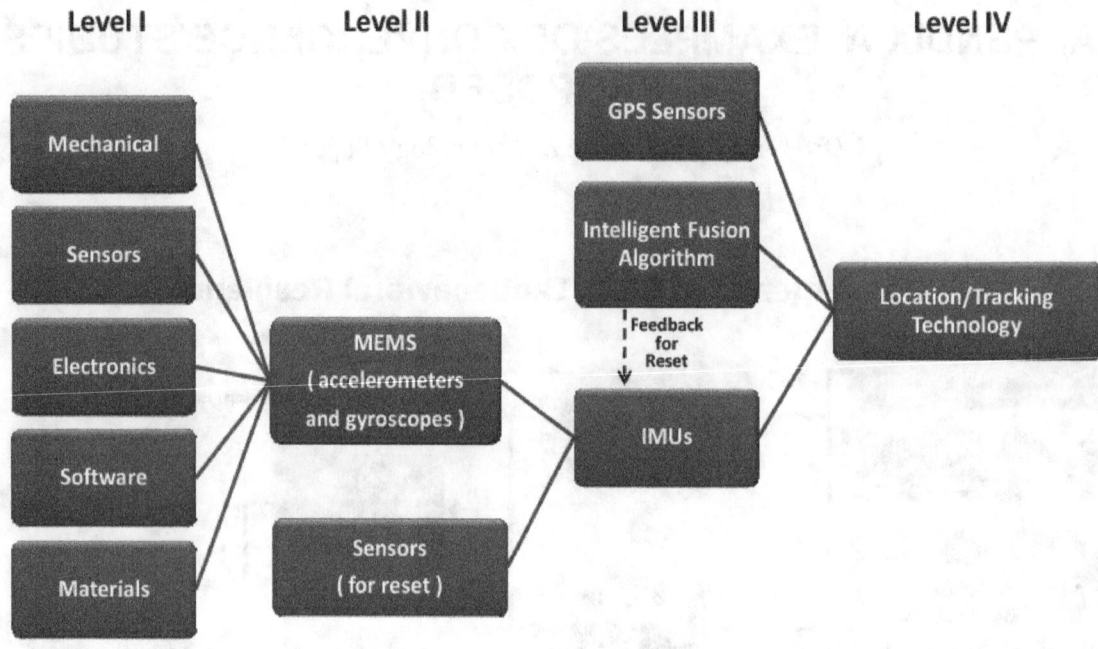

Figure 8. Roadmap for S&T Convergence in Materiel Location and Tracking Technology

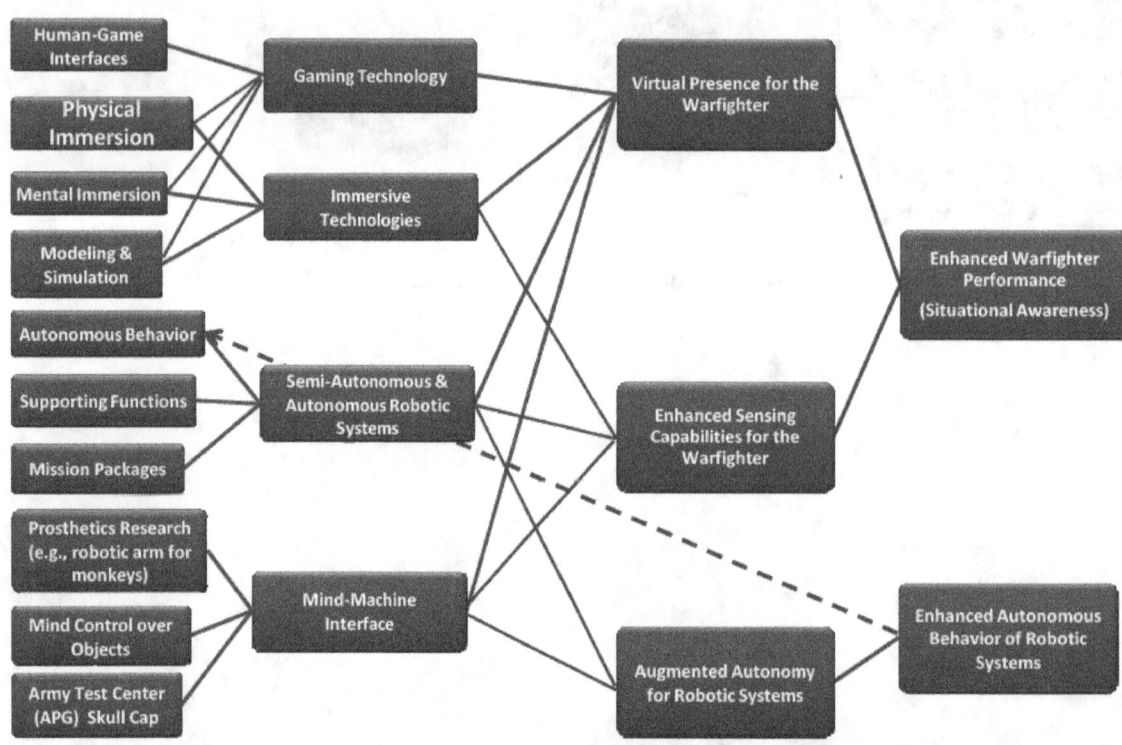

Figure 9. S&T Convergence (from left to right) to Enhance Warfighter Situational Awareness and the Autonomous Behavior of Robotic Systems

Convergence of S&T: Mechanochemical Transduction

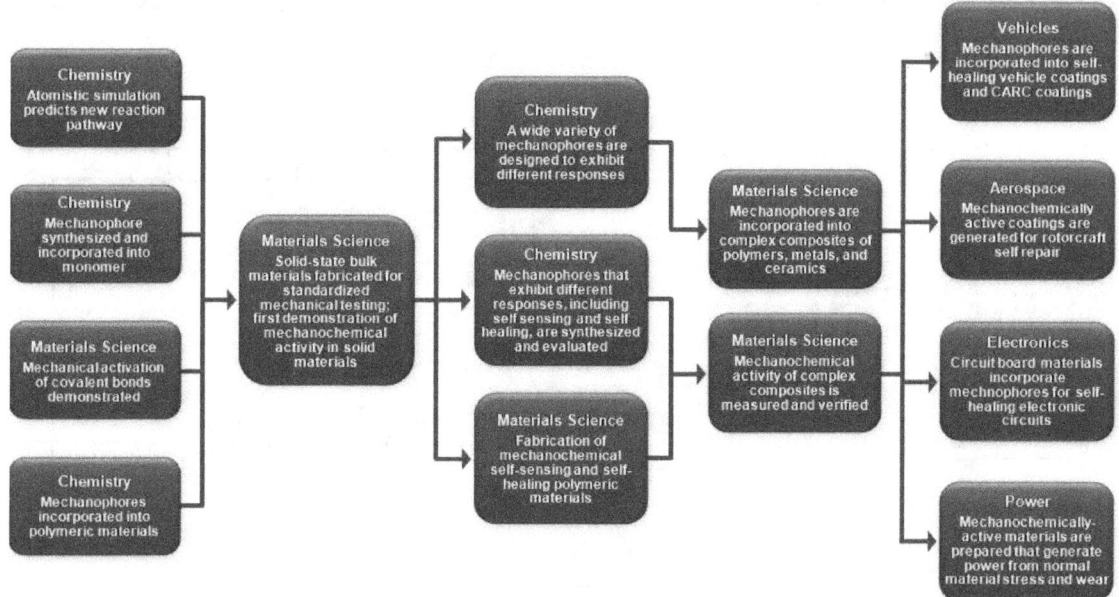

Figure 10. Convergence of S&T: Mechanochemical Transduction

APPENDIX B. RESULTS OF AN ACTUAL CONVERGENCE STUDY IN THE BIOLOGICAL SCIENCES (FROM REF. 27, DEFENSE & TECHNOLOGY PAPER 71)

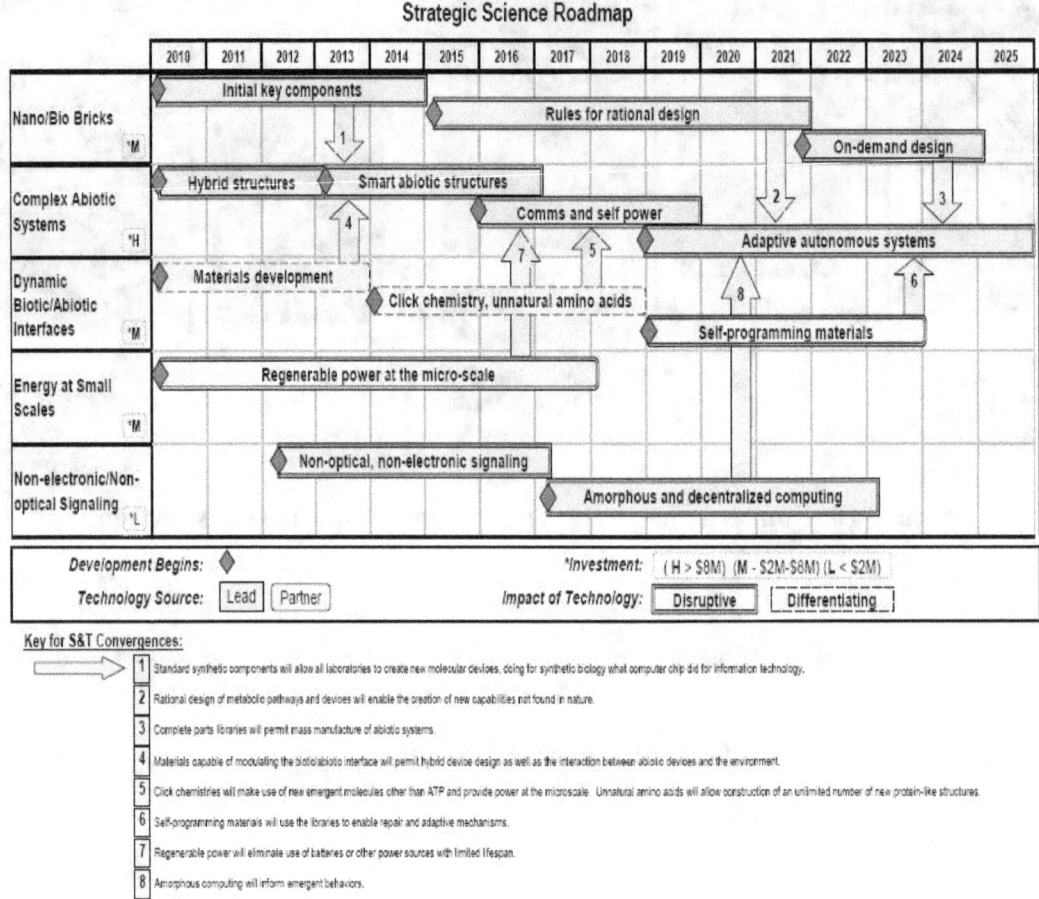

Strategic Science Roadmap

Key for S&T Convergences:

1 Standard synthetic components will allow all laboratories to create new molecular devices, doing for synthetic biology what computer chip did for information technology.

2 Rational design of metabolic pathways and devices will enable the creation of new capabilities not found in nature.

3 Complete parts libraries will permit mass manufacture of abiotic systems.

4 Materials capable of modulating the biotic/abiotic interface will permit hybrid device design as well as the interaction between abiotic devices and the environment.

5 Click chemistries will make use of new emergent molecules other than ATP and provide power at the microscale. Unnatural amino acids will allow construction of an unlimited number of new protein-like structures.

6 Self-programming materials will use the libraries to enable repair and adaptive mechanisms.

7 Regenerable power will eliminate use of batteries or other power sources with limited lifespan.

8 Amorphous computing will inform emergent behaviors.

APPENDIX C. 2012 MECHANOCHEMICAL TRANSDUCTION CONVERGENCE WORKSHOP

In January 2012, the Army Research Office held a Mechanochemical Transduction Convergence Workshop as a test case for identifying convergences of disciplines and the potential impact on science and the Army. A diverse group of academic and government scientists were invited to participate in the workshop. The 11 academic researchers were renowned subject matter experts representing a wide range of disciplines, including chemical engineering, organic chemistry, physical organic chemistry, molecular biochemistry, materials engineering, multi-scale theory, metallurgy, and physics. The chief objective of the workshop was to identify the most promising research opportunities and interdisciplinary convergences that could lead the field of mechanochemical transduction in new directions with unexpected outcomes relevant to future Army needs. The workshop was organized into three breakout sessions scheduled across 2 days, with a focus on discussion rather than presentations.

The first breakout session called for the academic participants to identify and forecast which of their research interests could potentially converge with complementary research from the other academics. The government participants met separately to identify specific research and technology areas of interest to the Army that could be affected by research in mechanochemical transduction.

The second breakout session used the results from the first session to focus the research convergences in directions relevant to Army interests. In the second session, the groups comprised academic researchers and at least one government participant. The objective of this session was to identify the most promising opportunities for substantial convergence in mechanochemical transduction that could affect the Army's top 24 science and technology challenges, with an emphasis on force protection, engineered resilient systems, and lethal effects.

The third breakout session built on the results of the second session and focused each group on refining its results and adding detail to produce a flowchart illustrating how parallel research in specific scientific domains could converge to create new fields of study and unexpected results. ARO is preparing a detailed report discussing the results of this workshop.

APPENDIX D. BRIEF SUMMARY OF A GEDANKEN EXPERIMENT AT FORT BENNING ON NETWORK SCIENCE AND SMALL UNIT OPERATIONS

The following was prepared by Albert Sciarretta for an internal National Defense University weekly news summary.

On January 18–19, 2012, Senior Research F ellow Albert Sciarretta conducted a Gedanken (Thought) Experiment at Fort Benning, Georgia. Center for Technology and National Security Policy (CTNSP) research analysts Jam es Garcia and Kathleen Jocoy s upported the event. The experiment solicited feedback from 14 m id-/senior-level Infantry non-comm issioned officers (NCO) and 14 captains (11 Infantry and 3 Arm or). All had com bat tours as squad leaders , platoon sergeants, or platoon leaders; with some of the NCOs having as many as six tours in Iraq or Afghanistan. Scientists an d engineers from Army laboratories and engineering centers participated as subject matter experts and obser vers. The experiment focused on exploring how future Network Science (dynam ic communications, information, and socio-cognitive networks) capabilities could be used by sm all units (platoon and below) in offensive, defensive, and stability operations, as well as on illum inating critical issues (e.g., in formation overload and distraction/disruption to the comm and and control structure) that should be considered as these capabilities are developed. The highly succes sful event generated a trem endous amount of information from discussions and surveys. A CTNSP Defense & Technology Paper will be written to docum ent the output o f the event. It will ad dress the process of designing and conducting the Gedank en Experiment, as well as the technical output—capab ility needs; potential operational issues; pot ential tactics, techniques, and procedures; and recommendations for science and technology investments. The partic ipating scientists and engineers have already begun to use the experim ent results in their re spective programs. Hearing about the event, an Army Science Board study committee requested a report on the key findings.

www.ingramcontent.com/pod-product-compliance
Lightning Source LLC
Chambersburg PA
CBHW081359170526
45166CB00010B/3146